Earth's
SPHERES

Rebecca Woodbury, Ph.D., M.Ed.

Gravitas Publications Inc.

Earth's
SPHERES

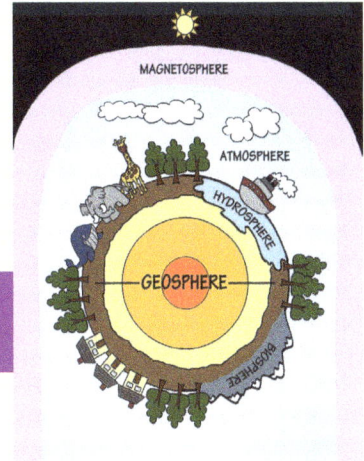

Illustrations: Janet Moneymaker

Earth's Spheres
ISBN 978-1-950415-37-3

Published by Gravitas Publications Inc.
Imprint: Real Science-4-Kids
www.gravitaspublications.com
www.realscience4kids.com

RS4K   Photo credits: Cover and Title Page: adimas, AdobeStock; P. 5 & 17, Photo of Earth, NASA

One way to learn about Earth
is to study Earth's spheres.

A **sphere** is an object that is shaped like a ball.

We can see that Earth is shaped like a ball when we look at it from space. So we can say that Earth is a sphere.

I didn't know that.

It's true!

Earth's spheres are the layers that surround Earth. We can study these spheres one at a time or together to learn how Earth works.

The **geosphere** is the part of Earth that is made of rocks and soils.

The rocky crust that surrounds Earth is part of the geosphere. The geosphere also includes the mantle and core, which are inside Earth.

THE GEOSPHERE

Crust The part of Earth where we live

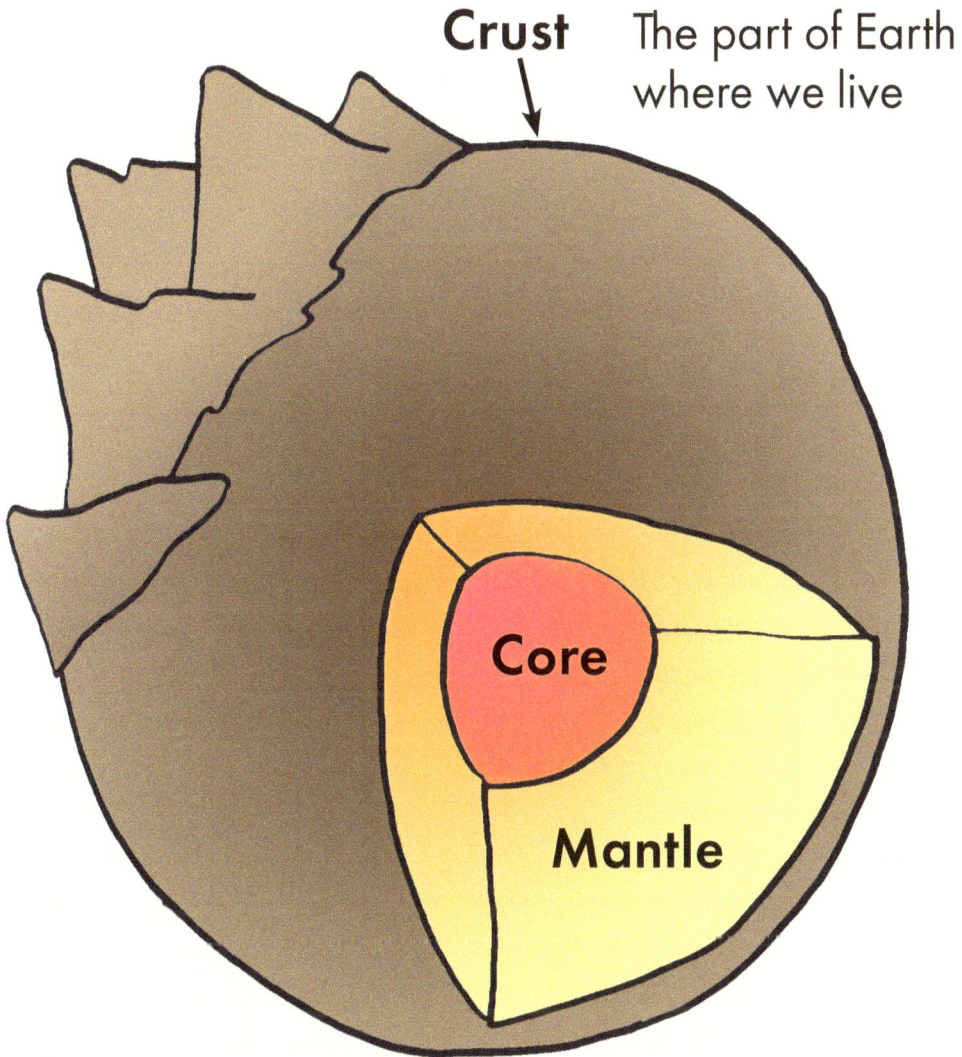

Core

Mantle

The **hydrosphere** is made up of all the water on Earth. The hydrosphere includes lakes, streams, oceans, ice, rain, and snow.

THE HYDROSPHERE

The **biosphere** is made up
of all the living things on Earth.

We are part of the biosphere!

Yes!

We are part of the biosphere too!

Yes!

THE BIOSPHERE

The **atmosphere** is made up of all the air that surrounds Earth. The air we breathe is in the atmosphere. Weather occurs in the atmosphere.

THE ATMOSPHERE

The **magnetosphere** is formed by the interaction of energy from the Sun with the magnetic field surrounding Earth. The magnetosphere protects Earth from getting too much energy from the Sun.

Sun

THE MAGNETOSPHERE

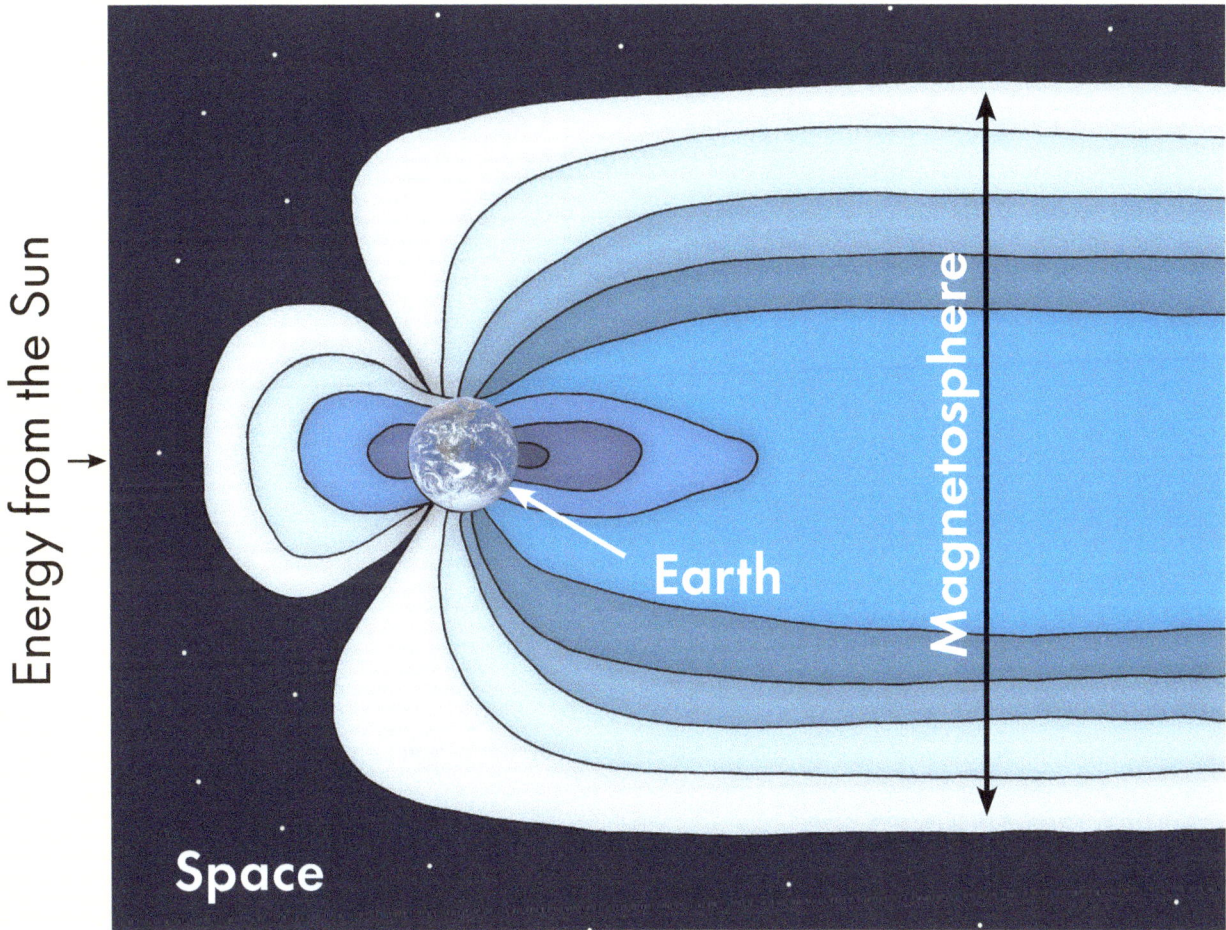

Energy from the Sun

Earth

Magnetosphere

Space

Each sphere plays an important part in making it possible for life to exist on Earth.

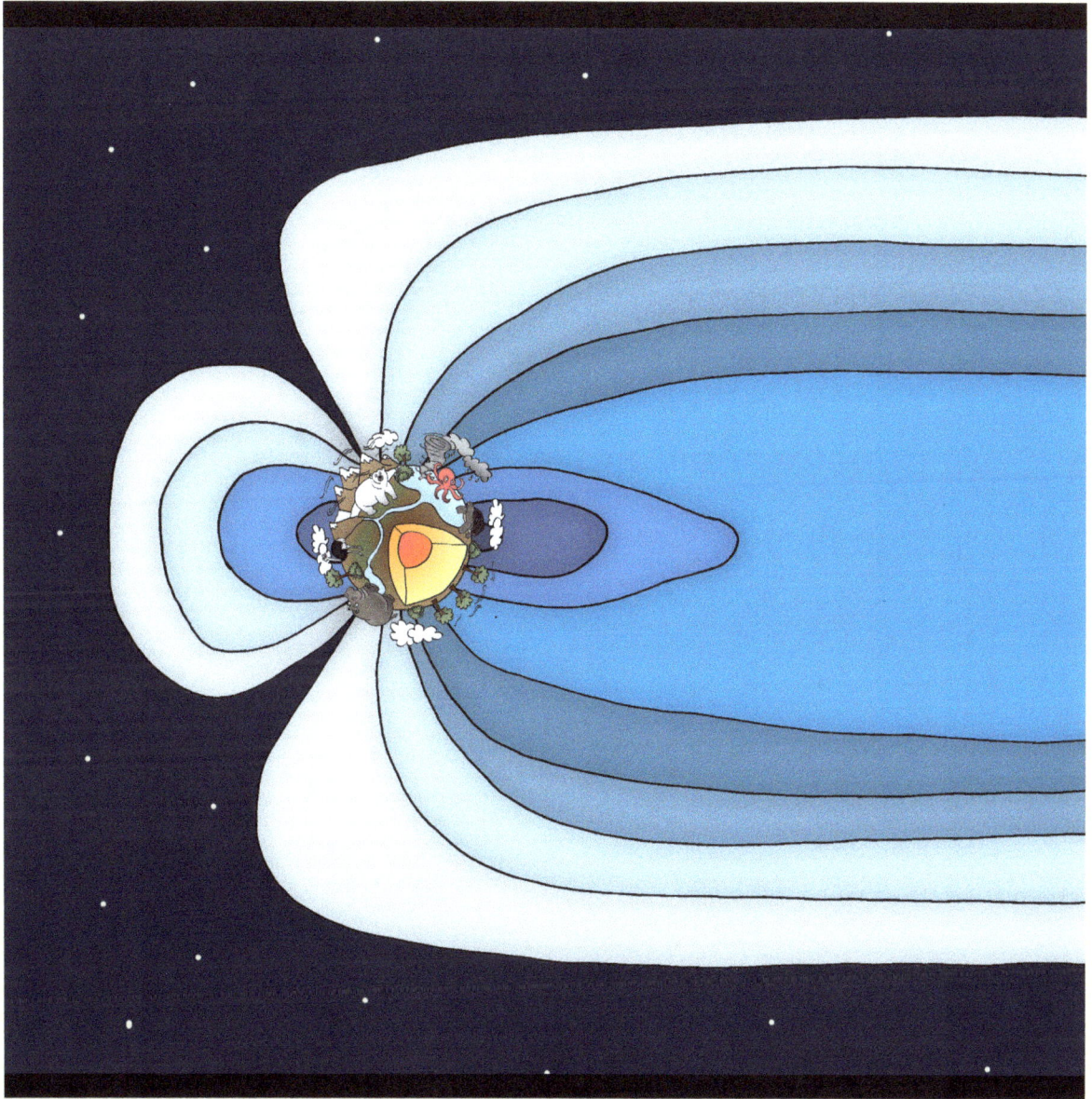

Together, the geosphere, hydrosphere, biosphere, atmosphere, and magnetosphere make Earth a special place that has everything we need to live.

Hooray for spheres!

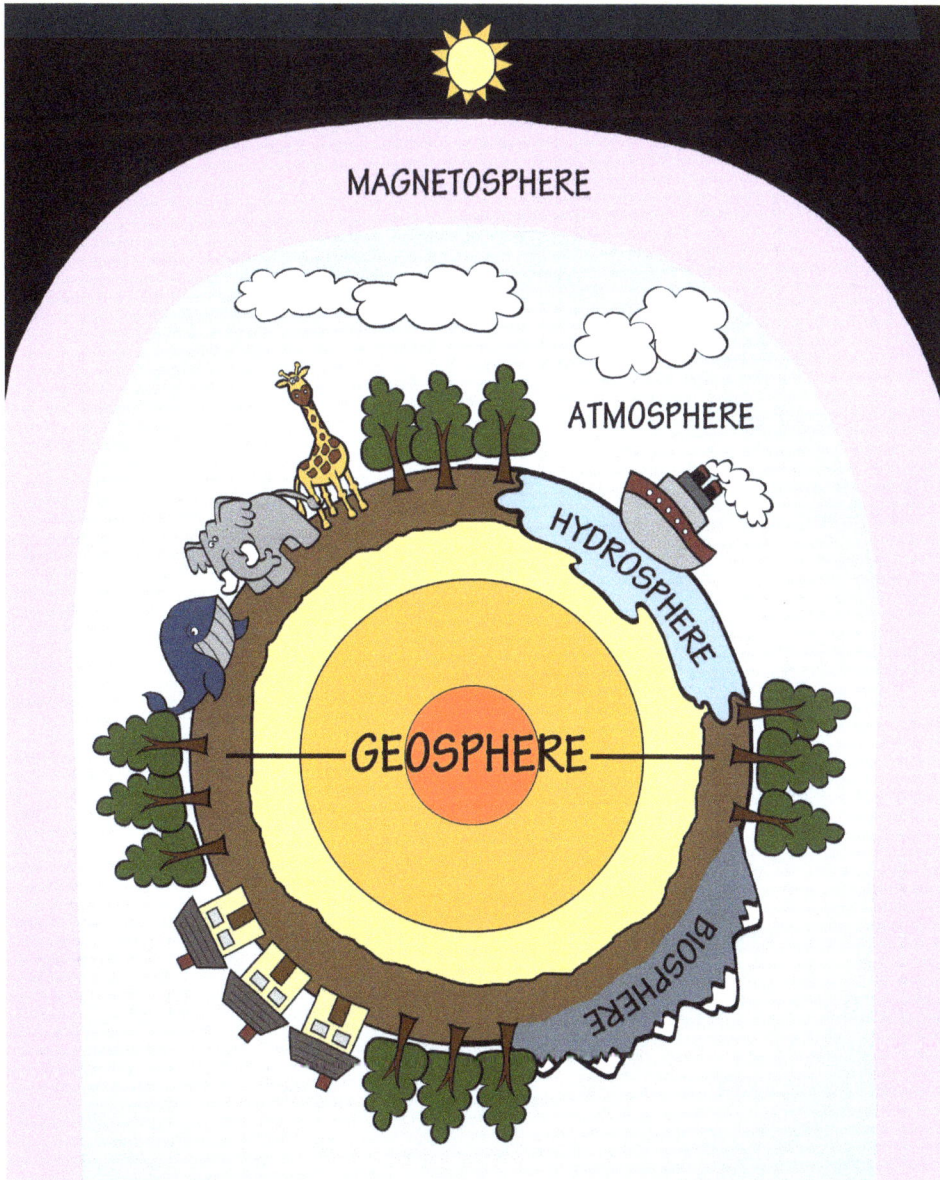

How to say science words

atmosphere (AAT-muh-sfeer)

biosphere (BIY-uh-sfeer)

core (KAWR)

crust (KRUHST)

energy (EH-nuhr-jee)

geosphere (JEE-oh-sfeer)

geometry (jee-AH-muh-tree)

hydrosphere (HIY-droh-sfeer)

layer (LAY-uhr)

magnetic field (maag-NEH-tik FEELD)

magnetosphere (maag-NEE-tuh-sfeer)

mantle (MAAN-tuhl)

science (SIY-uhns)

sphere (SFEER)

www.ingramcontent.com/pod-product-compliance
Lightning Source LLC
Chambersburg PA
CBHW040149200326
41520CB00028B/7545